MW00934739

EXPONENTS

ALGEBRA

EXPONENTS PRACTICE

PROBLEMS

WITH ANSWER KEY AND

STEP BY STEP SOLUTIONS

Copyright © 2019 N. Hirn

HOW TO USE THIS WORKBOOK

The study of Mathematics requires understanding of the concepts taught as well as practicing what is learned. It is advisable to practice mathematical problems using pencil and paper to allow the student to follow their train of thoughts as they write each step for the solutions.

This workbook provides problems that require the knowledge and use of the properties of EXPONENTS. These properties are summarized in the next section as a refresher for students.

When working out each problem, it is important to learn, understand and apply the correct property of EXPPONENTS in order to evaluate the mathematical expressions correctly.

Hints are provided when a problem requires the use of concepts that are not part of this workbook. Students should be familiar with some of these concepts.

The workbook is divided into four sections:

1. The first section includes a detail explanation of the topic discussed in the workbook with examples illustrating various problems and how each problem is being solved.

2. The second section includes a problem per page with ample lined space for the student to solve the problem. Using a pencil will allow the student to erase and redo steps as they become necessary.

3. The third section is the answer key.

4. The fourth section includes detailed solutions for each problem explaining the steps involved in reaching the answer.

In order to succeed the student should master the material in this section and practice solving each problem.

PROPERTIES OF EXPONENTS

When working on a mathematical expression it is important to understand how to evaluate each expression in order to obtain the correct answer.

In order to learn about EXPONENTS, it is important to first understand how to identify them and what they mean. This knowledge can then be transferred when learning to evaluate the properties associated with Exponents and how to use them to solve a problem.

Explanation

An "Exponent" is a number written just above and to the right of another number or variable which is called a "base".

An Exponent is used to show repeated multiplication of a certain base.

An Exponent is also called a Power.

A mathematical expression that involves an exponent is written in the following format: b^n

Where b is the called the base and n is the Exponent which represents the number of multiplications that the base undergoes. This means that the base b is multiplied by itself n times depending on what number "n" is.

Example: 3^2 The Base is 3.   The Exponent is 2 .

In order to evaluate the above expression, the base 3 must be multiplied by itself twice to get the answer as follows: $3^2 = 3 \cdot 3 = 9$

Hint:

A "base" can be a number or a variable. The following example shows this:

$$x^2$$

The above expression is evaluated the same way as if the base was a number. Therefore, the variable x is multiplied by itself twice.

Properties of Exponents:

Property   (1)

When multiplying two or more mathematical expressions with the same base, the exponents are added while keeping the same base.

Example: $3^2 \cdot 3^4 = 3^{2+4} = 3^6 = 729$

The above property cannot be applied if the bases are different numbers. When the bases are different, each base will have to be   evaluated separately before the operation is completed.

Example: $2^2 \cdot 3^4 = 4.81 = 324$

The above property applies to variable bases also.

Example: $x^2 \cdot x^4 = x^{2+4} = x^6$

The above property cannot be applied if the bases are different variables. When the bases are different variables, each base will have to be evaluated separately before the operation is completed.

Example: $x^2 \cdot y^4 = x^2 \, y^4$

Hint:

It is important to notice the following when evaluating bases with different variables:

Example: $x^2 \cdot y^4 x^3 = x^{2+3} \, y^4 = x^5 y^4$

As seen in the above example, the exponents are added only for the base x variable and not for the base y variable.

Property   (2)

When an expression (base) containing an exponent raised to another power, that base is raised to the product of the powers.

Example: $(3^2)^3 = 3^{2.3} = 3^6$

The above property applies to variable bases also.

Example: $(x^2)^3 = x^{2.3} = x^6$

Property   (3)

The product of two numbers or variables that are raised to an exponent (power) will result in the exponent being distributed over the product of the expression.

Example: $(2x)^3 = 2^3 \cdot x^3 = 8x^3$

The above property applies to variable bases also.

Example: $(yx)^3 = y^3.x^3 = y^3\,x^3$

Property   (4)

When dividing two or more mathematical expressions with the same   base, the exponent in the denominator is subtracted from the exponent in the numerator and the base is raised to the exponent that results.

Example: $\dfrac{2^5}{2^2} = 2^{5-2} = 2^3$

The above property applies to variable bases also.

Example: $\dfrac{x^5}{x^2} = x^{5-2} = x^3$

Property   (5)

When a quotient (two integers divided) is raised to an exponent or a power, each of the integers is each raised to the power in the result.

Example: $\left(\dfrac{x}{2}\right)^3 = \dfrac{x^3}{2^3} = \dfrac{x^3}{8}$

Hints to remember:

1.   *A base (whether a number or a variable) raised to the power of 1 equals itself.*

Example: $x^1 = x$

2. *A base raised to a negative exponent will always result in the reciprocal of the base raise to the positive exponent.*

   *It is important to remember that the base must not equal **0** since a division by **0** results in an undefined answer.*

   Example: $2^{-3} = \dfrac{1}{2^3} = \dfrac{1}{8}$

   The above applies to variable bases also.

   Example: $x^{-3} = \dfrac{1}{x^3}$

3. *A base (whether a number or a variable) raised to the power of **0** equals **1**.*

   Example: $2^0 = 1$

   The above applies to variable bases also.

   Example: $x^0 = 1$

4. *A base of **1** raised to any power equals itself.*

   Example: $1^3 = 1$

5. *Positive and negative signs included within a Mathematical expression are raised to the exponent (power) that the expression is raised to also.*

   It is important to remember how to multiply signs when working on expressions that include signs.

   Example: $-3^2 = (-3)(-3) = 9$

   The above applies to variable bases also.

   Example: $-x^2 = (-x)(-x) = +x$

PROBLEM # 1

SOLVE: 4^2

YOUR WORK:

_____

_____

_____

_____

_____

_____

_____

_____

_____

_____

_____

_____

_____

_____

_____

_____

PROBLEM # 2

SOLVE: $\left(-\dfrac{5}{6}\right)^2$ and $\left(-\dfrac{5}{6}\right)^3$

YOUR WORK:

_____

_____

_____

_____

_____

_____

_____

_____

_____

_____

_____

_____

_____

_____

_____

_____

_____

PROBLEM # 3

SOLVE: $(2^3)^2$

YOUR WORK:

PROBLEM # 4

SOVE: -2^3

YOUR WORK:

_____

_____

_____

_____

_____

_____

_____

_____

_____

_____

_____

_____

_____

_____

_____

_____

_____

PROBLEM # 5

SOLVE: $\left(\dfrac{2}{3}\right)^2$ and $\dfrac{2^2}{3}$

YOUR WORK:

_____

_____

_____

_____

_____

_____

_____

_____

_____

_____

_____

_____

_____

_____

_____

_____

_____

_____

PROBLEM # 6

SOLVE: $\left(-\frac{2}{3}x^2\right)^3$

YOUR WORK:

PROBLEM # 7

SOLVE: $x^4 \cdot x^5$

YOUR WORK:

_____

_____

_____

_____

_____

_____

_____

_____

_____

_____

_____

_____

_____

_____

_____

_____

_____

PROBLEM # 8

SOLVE: $6x^2(-3x^4)(2x^5)$

YOUR WORK:

PROBLEM # 9

SOLVE: $\left(-\frac{1}{3}\,n\right)^4 (2n^3)^2 \left(\frac{3}{2}\,n^6\right)^4$

YOUR WORK:

_____

_____

_____

_____

_____

_____

_____

_____

_____

_____

_____

_____

_____

_____

_____

_____

PROBLEM # 10

SOLVE: $(5y^4)^{-3}(2y^{-2})^3$

YOUR WORK:

PROBLEM # 11

SOLVE: $x^4 \cdot x^6 \cdot x^8 \cdot x^{10}$

YOUR WORK:

PROBLEM # 12

SOLVE: $\left(\frac{1}{2} x^3\right)\left(\frac{2}{3} x^4\right)\left(\frac{3}{5} x^{-7}\right)$

YOUR WORK:

_____

_____

_____

_____

_____

_____

_____

_____

_____

_____

_____

_____

_____

_____

_____

_____

PROBLEM # 13

SOLVE: $\dfrac{5^3}{5^1}$

YOUR WORK:

_____

_____

_____

_____

_____

_____

_____

_____

_____

_____

_____

_____

_____

_____

_____

_____

PROBLEM # 14

SOLVE: $\left(4a^5\, b^2\right)\left(2b^{-5}\, c^2\right)\left(3a^7\, c^4\right)$

YOUR WORK:

_____

_____

_____

_____

_____

_____

_____

_____

_____

_____

_____

_____

_____

_____

_____

_____

PROBLEM # 15

SOLVE: $(-3x)^4$

YOUR WORK:

PROBLEM # 16

SOLVE: $\dfrac{(m^3)^2 m^5}{(m^4)^3}$

YOUR WORK:

PROBLEM # 17

SOLVE: $(2x^4)^3$

YOUR WORK:

PROBLEM # 18

SOLVE: $\dfrac{\left(x^{-2}\right)^3 \left(x^3\right)^{-2}}{x^{10}}$

YOUR WORK:

PROBLEM # 19

SOLVE: $(3x^2)^3 (2x)^4$

YOUR WORK:

_____

_____

_____

_____

_____

_____

_____

_____

_____

_____

_____

_____

_____

_____

_____

_____

_____

PROBLEM # 20

SOLVE: $\dfrac{27x^3y^{-4}z}{9x^7y^{-6}z^4}$

YOUR WORK:

_____

_____

_____

_____

_____

_____

_____

_____

_____

_____

_____

_____

_____

_____

_____

_____

_____

_____

PROBLEM # 21

SOLVE: 3^{-2}

YOUR WORK:

_____

_____

_____

_____

_____

_____

_____

_____

_____

_____

_____

_____

_____

_____

_____

_____

_____

PROBLEM # 22

SOLVE: $(5y)^{-2}$

YOUR WORK:

PROBLEM # 23

SOLVE: $\dfrac{x^{10}}{x^4}$

YOUR WORK:

_____

_____

_____

_____

_____

_____

_____

_____

_____

_____

_____

_____

_____

_____

_____

_____

SOLVE: $\dfrac{\left(x^0\right)^5}{x^3}$

YOUR WORK:

_____

_____

_____

_____

_____

_____

_____

_____

_____

_____

_____

_____

_____

_____

_____

_____

_____

PROBLEM # 25

SOLVE: $\dfrac{12r^{-6}s^0t^{-3}}{3\,r^{-4}s^{-3}t^{-5}}$

YOUR WORK:

PROBLEM # 26

SOLVE: $\dfrac{\left(3x^{-2}y^8\right)^4}{\left(9x^4\,y^{-3}\right)^2}$

YOUR WORK:

PROBLEM # 27

SOLVE: $(3x^2)(2x^3)(5x^4)$

YOUR WORK:

SOLVE: $\left(\dfrac{8x^2}{4x^4\,y^{-3}}\right)^4$

YOUR WORK:

_____

_____

_____

_____

_____

_____

_____

_____

_____

_____

_____

_____

_____

_____

_____

_____

_____

_____

SOLVE: $\dfrac{2^{-5}}{2^3}$

YOUR WORK:

_____

_____

_____

_____

_____

_____

_____

_____

_____

_____

_____

_____

_____

_____

_____

_____

SOLVE: $\left(2x^2y^{-5}\right)^3\left(3x^{-4}y^2\right)^{-4}$

YOUR WORK:

_____

_____

_____

_____

_____

_____

_____

_____

_____

_____

_____

_____

_____

_____

_____

_____

PROBLEM # 31

SOLVE: $\dfrac{t^{-10}}{t^{-4}}$

YOUR WORK:

_____

_____

_____

_____

_____

_____

_____

_____

_____

_____

_____

_____

_____

_____

_____

_____

_____

SOLVE: $\dfrac{(3x)^{-5}}{(3x)^{-8}}$

YOUR WORK:

PROBLEM # 33

SOLVE: $(2x^4 y^{-3})(7x^{-8} y^5)$

YOUR WORK:

_____

_____

_____

_____

_____

_____

_____

_____

_____

_____

_____

_____

_____

_____

_____

_____

PROBLEM # 34

SOLVE: $(2a^2b)^0$

YOUR WORK:

PROBLEM # 35

SOLVE: $\dfrac{\left(2a^3b^{-2}c\right)^5}{\left(a^{-2}b^4c^{-3}\right)^{-2}}$

YOUR WORK:

_____

_____

_____

_____

_____

_____

_____

_____

_____

_____

_____

_____

_____

_____

_____

_____

PROBLEM # 36

SOLVE: $\dfrac{\left(a^3\right)^2 \left(a^4\right)^5}{\left(a^5\right)^2}$

YOUR WORK:

_____

_____

_____

_____

_____

_____

_____

_____

_____

_____

_____

_____

_____

_____

_____

_____

SOLVE: $2x^2y^3\left(\dfrac{7xy^4}{14x^3y^6}\right)^{-2}$

YOUR WORK:

_____

_____

_____

_____

_____

_____

_____

_____

_____

_____

_____

_____

_____

_____

_____

_____

_____

PROBLEM # 38

SOLVE: $\dfrac{\left(x^{-7}\right)^3 \left(x^4\right)^5}{\left(x^3\right)^2 \left(x^{-1}\right)^8}$

YOUR WORK:

_____

_____

_____

_____

_____

_____

_____

_____

_____

_____

_____

_____

_____

_____

_____

_____

PROBLEM # 39

SOLVE: $\left(\dfrac{x^{-5}y^2}{x^{-3}y^5} \right)^{-2}$

YOUR WORK:

_____

_____

_____

_____

_____

_____

_____

_____

_____

_____

_____

_____

_____

_____

_____

_____

_____

PROBLEM # 40

SOLVE: $\dfrac{\left(a^{-2}\right)^3 \left(a^4\right)^2}{\left(a^{-3}\right)^{-2}}$

YOUR WORK:

PROBLEM # 41

SOLVE: $8x^4y^{-3}\left(\dfrac{12x^{-3}y^{-2}}{24x^4y^{-5}}\right)^0$

YOUR WORK:

_____

_____

_____

_____

_____

_____

_____

_____

_____

_____

_____

_____

_____

_____

_____

_____

_____

_____

PROBLEM # 42

SOLVE: $\left(\dfrac{ab^{-3}c^{-2}}{a^{-3}b^0c^{-5}}\right)^{-1}$

YOUR WORK:

PROBLEM # 43

SOLVE: $\dfrac{5a^8b^3}{20a^5b^{-4}}$

YOUR WORK:

_____

_____

_____

_____

_____

_____

_____

_____

_____

_____

_____

_____

_____

_____

_____

_____

_____

SOLVE: $\left(\dfrac{x^{-3}y^2}{x^4y^{-5}}\right)^{-2}\left(\dfrac{x^{-4}y^2}{x^0y^{-3}}\right)$

YOUR WORK:

PROBLEM # 45

SOLVE: $\dfrac{\left(x^5\right)^6}{\left(x^3\right)^4}$

YOUR WORK:

ANSWER KEY

1. **16**

2. $\left(\frac{25}{36}\right)$ & $\left(-\frac{25}{36}\right)$

3. **64**

4. -8

5. $\left(\frac{24}{9}\right)$ & $\left(\frac{4}{3}\right)$

6. $-\frac{8}{27}\,x^6$

7. x^9

8. $-36\,x^{11}$

9. $\frac{1}{4}\text{n}^{34}$

10. $\left(\frac{8}{125y^{18}}\right)$

11. x^{28}

12. $\frac{1}{5}$

13. **25**

14. $\frac{24a^{12}c^6}{b^3}$

15. $81\,x^4$

16. $\dfrac{1}{m}$

17. $8x^{12}$

18. $\dfrac{1}{x^{22}}$

19. $432\,x^{10}$

20. $\dfrac{3y^2}{x^4 z^3}$

21. $\dfrac{1}{9}$

22. $\dfrac{1}{25y^2}$

23. x^6

24. $\dfrac{1}{x^3}$

25. $\dfrac{4\,s^3 t^2}{r^2}$

26. $\dfrac{y^{38}}{x^{16}}$

27. $30x^9$

28. $\dfrac{16y^{12}}{x^8}$

29. $\dfrac{1}{2^8}$

30. $\dfrac{8x^{22}}{81y^{23}}$

31. $\dfrac{1}{t^6}$

32. $27x^3$

33. $\dfrac{14y^2}{x^4}$

34. 1

35. $\dfrac{32a^{11}}{b^2c}$

36. a^{16}

37. $8\,x^6\,y^7$

38. x

39. x^4y^6

40. $\dfrac{1}{a^4}$

41. $\dfrac{8x^4}{y^3}$

42. $\dfrac{b^3}{a^4c^3}$

43. $\dfrac{a^3\,b^7}{4}$

44. $\dfrac{x^{10}}{y^{15}}$

45. x^{18}

PROBLEM # 1

SOLVE: 4^2

CORRECT ANSWER: 16

In the above expression, the exponent is **(2)** and the base is **(4)**. This means that the base **(4)** is multiplied by itself twice as follows:
$$4^2 = 4 \cdot 4 = 16$$

PROBLEM # 2

SOLVE: $\left(-\dfrac{5}{6}\right)^2$ **and** $\left(-\dfrac{5}{6}\right)^3$

CORRECT ANSWER: $\left(\dfrac{25}{36}\right)$ **and** $\left(-\dfrac{25}{36}\right)$

The above expression shows two identical fractions $\left(-\dfrac{5}{6}\right)$. This fraction is the base which includes the fraction and its sign. The sign is included with the number because they are both enclosed within a parenthesis.
The first base has an even exponent which is **(2)**.

The second fraction has an odd exponent which is **(3)**.

The purpose of this exercise is to show how the negative sign is affected with the use of an even versus an odd exponent.

Also:

The above expression follows property (5) of the exponent rules which are explained in the front of this workbook.

Hint 1:

Multiplying a negative sign that is raised to an even exponent always results in a positive answer.

$$\left(-\frac{5}{6}\right)^2 = \left(-\frac{5}{6}\right)\left(-\frac{5}{6}\right) = \left(+\frac{25}{36}\right) = \left(\frac{25}{36}\right)$$

Hint 2:

Multiplying a negative sign that is raised to an odd exponent always results in a negative answer.

$$\left(-\frac{5}{6}\right)^3 = \left(-\frac{5}{6}\right)\left(-\frac{5}{6}\right)\left(-\frac{5}{6}\right) = \left(-\frac{25}{36}\right)$$

PROBLEM # 3

SOLVE: $(2^3)^2$

CORRECT ANSWER: 64

The above expression follows property (2) of the exponent rules which are explained in the front of this workbook.

Follow property (2) and simplify the problem by raising what is inside each parenthesis to the respected exponent.
$(2^3)^2 = (2)^{3 \cdot 2} = 2^6 = 64$

PROBLEM # 4

SOLVE: -2^3

CORRECT ANSWER: -8

In the above expression the base is **(2).** The exponent is **(3).**

It is important to note that the negative sign is not part of the base when evaluating this expression. This is because it is not enclosed within a parenthesis. Therefore, when doing the base multiplication, the negative sign is not multiplied. Only the number **(2)** is multiplied. This will result in a negative answer even though the exponent is odd.

Here is how to work this problem: $-2^3 = -2 \cdot 2 \cdot 2 = -8$

PROBLEM # 5

SOLVE: $\left(\dfrac{2}{3}\right)^2$ and $\dfrac{2^2}{3}$

CORRECT ANSWER: $\dfrac{4}{9}$ & $\dfrac{4}{3}$

The above expression shows two fractions; both are $\dfrac{2}{3}$. The difference is that the first fraction is enclosed within a parenthesis making it the base raised to the second power. In the second fraction, the numerator **(2)** is the base since it is raised to the second power and not the denominator **(3)**. Also:

The first fraction follows property (5) of the exponent rules which are explained in the front of this workbook.

Initially follow property (5) and simplify the problem by raising each integer inside the parenthesis to the respected exponent as follows:

$$\left(\dfrac{2}{3}\right)^2 = \left(\dfrac{2}{3}\right)\left(\dfrac{2}{3}\right) = \dfrac{4}{9}$$

Here is how you solve the second fraction: $\dfrac{2^2}{3} = \dfrac{2 \cdot 2}{3} = \dfrac{4}{3}$

PROBLEM # 6

SOLVE: $\left(-\dfrac{2}{3}x^2\right)^3$

CORRECT ANSWER: $-\dfrac{8}{27}x^6$

The above expression follows property (2) of the exponent rules which are explained in the front of this workbook.

Initially follow property (2) and simplify the problem by raising what is inside the parenthesis to the respected exponent as follows: $\left(-\dfrac{2}{3}\right)^3 (x^2)^3$

Here is how you solve the problem:

$$\left(-\frac{2}{3}\right)^3 (x^2)^3 = \left(-\frac{2}{3}\right)\left(-\frac{2}{3}\right)\left(-\frac{2}{3}\right)(x)^{2\cdot3} = -\frac{8}{27} x^6$$

PROBLEM # 7

SOLVE: $x^4 \cdot x^5$

CORRECT ANSWER: x^9

The above expression follows property (1) of the exponent rules which are explained in the front of this workbook.

Here is how you solve the problem: $x^4 \cdot x^5 = x^{4+5} = x^9$

PROBLEM # 8

SOLVE: $6x^2(-3x^4)(2x^5)$

CORRECT ANSWER: $-36\, x^{11}$

The above expression follows property (1) of the exponent rules which are explained in the front of this workbook.

Initially follow property (1) and multiply the numbers and adding the exponents on the variable as follows:

$$6x^2(-3x^4)(2x^5) = \quad (6)(-3)(2)x^{2+4+5} = -36\, x^{11}$$

PROBLEM # 9

SOLVE: $\left(-\frac{1}{3} n\right)^4 (2n^3)^2 \left(\frac{3}{2} n^6\right)^4$

CORRECT ANSWER: $\frac{1}{4}n^{34}$

The above expression follows properties (1) and (2) and (3) of the exponent rules that are explained in the front of this workbook.

Initially follow properties (2) and (3) and rewrite the problem to simplify solving it as follows:

$$\left(-\frac{1}{3}\right)^4 n^4 (2)^2 n^{3\cdot 2} \left(\frac{3}{2}\right)^4 n^{6\cdot 4}$$

Next multiply the exponents on the variable as seen below:

$$\left(-\frac{1}{3}\right)^4 n^4 (2)^2 n^6 \left(\frac{3}{2}\right)^4 n^{24}$$

Next raise the bases to the powers and group numbers and variables as follows:

$$\left(-\frac{1}{3}\right)\left(-\frac{1}{3}\right)\left(-\frac{1}{3}\right)\left(-\frac{1}{3}\right) (2)(2) \left(\frac{3}{2}\right)\left(\frac{3}{2}\right)\left(\frac{3}{2}\right)\left(\frac{3}{2}\right) n^4 n^6 n^{24}$$

Hint:

*The number **(2)** above is actually a fraction $\frac{2}{1}$*

$$\left(-\frac{1}{3}\right)\left(-\frac{1}{3}\right)\left(-\frac{1}{3}\right)\left(-\frac{1}{3}\right) \left(\frac{2}{1}\right)\left(\frac{2}{1}\right) \left(\frac{3}{2}\right)\left(\frac{3}{2}\right)\left(\frac{3}{2}\right)\left(\frac{3}{2}\right) n^4 n^6 n^{24}$$

Next, multiply the numbers while adding the exponents of the variables.

$$\left(\frac{1}{81}\right)\left(\frac{4}{1}\right)\left(\frac{81}{16}\right) n^{4+6+24} = \left(\frac{1}{81}\right)\left(\frac{4}{1}\right)\left(\frac{81}{16}\right) n^{34}$$

The fractions can be simplified and the final result is: $\frac{1}{4}n^{34}$

PROBLEM # 10

SOLVE: $(5y^4)^{-3}(2y^{-2})^3$

CORRECT ANSWER: $\left(\frac{8}{125y^{18}}\right)$

The above expression follows properties (1) and (2) of the exponent rules which are explained in the front of this workbook.

And also:

A base raised to a negative exponent will always result in the reciprocal of the base raise to the positive exponent.

Initially follow property (2) and simplify the problem by raising what is inside each parenthesis to the respected exponent.

$$(5^{(-3)}y^{4\cdot(-3)})(2^3y^{-2\cdot3}) = \quad (5^{-3}y^{-12})(2^3y^{-6})$$

Rewrite the problem with positive exponents by using the reciprocal of the base raised to a positive exponent as follows: $\left(\dfrac{1}{5^3}\cdot\dfrac{1}{y^{12}}\right)\left(\dfrac{2^3}{1}\cdot\dfrac{1}{y^6}\right)$

Perform the multiplication operation as follows: $\left(\dfrac{1}{5^3y^{12}}\right)\left(\dfrac{2^3}{y^6}\right)$

Perform the multiplication operation next. Keep in mind that multiplying the (y) variable follows property (1) that requires the exponents (powers) to be added: $\left(\dfrac{2^3}{5^3y^{12+6}}\right) = \left(\dfrac{2^3}{5^3y^{18}}\right)$

Finally raise the numbers to the related exponents as follows: $\left(\dfrac{8}{125y^{18}}\right)$

PROBLEM # 11

SOLVE: $\quad x^4 \cdot x^6 \cdot x^8 \cdot x^{10}$

CORRECT ANSWER: x^{28}

The above expression follows property (1) of the exponent rules which are explained in the front of this workbook.

Initially follow property (1) and add the exponent on the variable as follows:

$$x^{4+6+8+10} = x^{28}$$

PROBLEM # 12

SOLVE: $\left(\dfrac{1}{2}x^3\right)\left(\dfrac{2}{3}x^4\right)\left(\dfrac{3}{5}x^{-7}\right)$

CORRECT ANSWER: $\dfrac{1}{5}$

The above expression follows property (1) of the exponent rules which are explained in the front of this workbook.

Initially follow property (1) by regrouping the numbers and adding the exponents on the variable as follows:

$$\left(\frac{1}{2}\right)\left(\frac{2}{3}\right)\left(\frac{3}{5}\right)x^{3+4+(-7)} = \left(\frac{6}{30}\right)x^0$$

Based on the information explained in the front of this workbook, any base raised to a zero power will equal **(1)**, therefore $\left(\dfrac{6}{30}\right)\mathbf{1} = \dfrac{6}{30}$

This fraction can be simplified to: $\dfrac{1}{5}$

PROBLEM # 13

SOLVE: $\dfrac{5^3}{5^1}$

CORRECT ANSWER: 25

A base is raised to the first power will always equal itself therefore the **(5)** in the denominator raised to the power of **(1)** will equal **(5)**.

Multiply the numerator by itself three times to get: $\dfrac{125}{5}$

This can be simplified to: **25**

PROBLEM # 14

SOLVE: $\left(4a^5\,b^2\right)\left(2b^{-5}\,c^2\right)\left(3a^7\,c^4\right)$

CORRECT ANSWER: $\dfrac{24a^{12}c^6}{b^3}$

The above expression follows property (1) of the exponent rules which are explained in the front of this workbook.

And also:

A base raised to a negative exponent will always result in the reciprocal of the base raise to the positive exponent.

Initially follow property (1) by grouping the numbers and adding the exponents on the similar bases as follows: $(4)(2)(3)a^{5+7} b^{2+(-5)} c^{2+4}$

Multiply the numbers and add the exponents on the different variables as follows: $24 \, a^{12} b^{-3} c^{6}$

Rewrite the problem with positive exponents by using the reciprocal of the base (b) raised to a positive exponent as follows: $24 \, a^{12} \dfrac{1}{b^3} c^6$

Complete the problem by multiplying the numerators as follows. Remember that the number **(24)** and the variables (a) and (c) are each a fraction over the number one. This can be rewritten as follows: $\dfrac{24}{1} \dfrac{a^{12}}{1} \dfrac{1}{b^3} \dfrac{c^6}{1}$

Finally complete the final multiplication to get the answer: $\dfrac{24a^{12}c^6}{b^3}$

PROBLEM # 15

SOLVE: $(-3x)^4$

CORRECT ANSWER: $81 \, x^4$

The above expression shows the base to be $(-3x)$.

The sign is included with the number because they are both enclosed within a parenthesis.

The purpose of this exercise is to show how the negative sign is affected with the use of an even exponent.

Also:

The above expression follows property (2) of the exponent rules which are explained in the front of this workbook.

The problem can be written as follows to simplify it:

$$(-3)(-3)(-3)(-3)(x)(x)(x)(x) = \mathbf{81\ x^4}$$

Hint:

Multiplying a negative sign that is raised to an even exponent always results in a positive answer.

PROBLEM # 16

SOLVE: $\dfrac{\left(m^3\right)^2 m^5}{\left(m^4\right)^3}$

CORRECT ANSWER: $\dfrac{1}{m}$

The above expression follows properties (1), (2) and (4) of the exponent rules which are explained in the front of this workbook.

Initially follow property (2) and simplify the problem by raising what is inside each parenthesis to the respected exponent.

$$\frac{(m^{3\cdot2})m^5}{(m^{4\cdot3})} = \frac{m^6 m^5}{m^{12}}$$

Next follow property (1) and add the exponents on the same base in the numerator as follows: $\dfrac{m^{6+5}}{m^{12}} = \dfrac{m^{11}}{m^{12}}$

Next follow property (4) and subtract the exponent on the same base as follows: $m^{11-12} = m^{-1}$

Finally, rewrite the problem with positive exponents by using the reciprocal of the base (m) raised to a positive exponent as follows: $\dfrac{1}{m}$

PROBLEM # 17

SOLVE: $(2x^4)^3$

CORRECT ANSWER: $8x^{12}$

The above expression follows property (2) of the exponent rules which are explained in the front of this workbook.

Initially follow property (2) and simplify the problem by raising what is inside each parenthesis to the respected exponent as follows: $2^3 x^{4 \cdot 3} = 8x^{12}$

PROBLEM # 18

SOLVE: $\dfrac{(x^{-2})^3 (x^3)^{-2}}{x^{10}}$

CORRECT ANSWER: $\dfrac{1}{x^{22}}$

The above expression follows properties (1), (2) and (4) of the exponent rules which are explained in the front of this workbook.

And also:

A base raised to a negative exponent will always result in the reciprocal of the base raise to the positive exponent.

Initially follow property (2) and simplify the problem by raising what is inside each parenthesis to the respected exponent as follows:

$$\frac{x^{-2 \cdot 3} \ x^{3 \cdot -2}}{x^{10}} = \frac{x^{-6} \ x^{-6}}{x^{10}}$$

Next, follow property (2) and add the exponents on the base in the numerator as follows: $\dfrac{x^{-6 + -6}}{x^{10}} = \dfrac{x^{-12}}{x^{10}}$

Next, follow property (4) and subtract the exponents on the similar base in the fraction as follows: $x^{-12-10} = x^{-22}$

Finally, use the reciprocal of the answer in order to leave the solution in a positive exponent as follows: $\dfrac{1}{x^{22}}$

PROBLEM # 19

SOLVE: $(3x^2)^3 (2x)^4$

CORRECT ANSWER: $432\,x^{10}$

The above expression follows properties (1), (2) and (3) of the exponent rules which are explained in the front of this workbook.

Initially follow properties (2) and (3) and raise the bases inside the parenthesis to the respected exponents as follows:

$(3^3 x^{2\cdot 3})\,(2^4 x^4) = (3^3 x^6)\,(2^4 x^4)$

Next, raise the numbers to the exponents as follows: $(27x^6)\,(16x^4)$

Finally, follow property (1) by multiplying the numbers and adding the exponents on the variable to solve the problem as follows:

$432\,x^{6+4} = 432\,x^{10}$

PROBLEM # 20

SOLVE: $\dfrac{27x^3 y^{-4} z}{9x^7 y^{-6} z^4}$

CORRECT ANSWER: $\dfrac{3y^2}{x^4 z^3}$

The above expression follows property (4) of the exponent rules which are explained in the front of this workbook.

And also:

A base raised to a negative exponent will always result in the reciprocal of the base raise to the positive exponent.

Initially simplify the numbers by using the division operation as follows:

$27 \div 9 = 3$

Next follow property (4) and rewrite the problem in a simplified manner. as follows: $3x^{3-7}y^{-4-(-6)}z^{1-4} = 3x^{-4}y^{-4+6}z^{1-4} = 3x^{-4}y^2z^{-3}$

Finally, use the reciprocal of the answer in order to leave the solution in a positive exponent as follows: $\dfrac{3y^2}{x^4z^3}$

PROBLEM # 21

SOLVE: 3^{-2}

CORRECT ANSWER: $\dfrac{1}{9}$

A base raised to a negative exponent will always result in the reciprocal of the base raise to the positive exponent.

Uses the reciprocal of the base raised to a positive exponent and solve the problem as follows: $\dfrac{1}{3^2} = \dfrac{1}{9}$

PROBLEM # 22

SOLVE: $(5y)^{-2}$

CORRECT ANSWER: $\dfrac{1}{25y^2}$

The above expression follows property (3) of the exponent rules which are explained in the front of this workbook.

And also:

A base raised to a negative exponent will always result in the reciprocal of the base raise to the positive exponent.

Initially follow property (3) and raise the base inside the parenthesis to the exponent as follows: $5^{-2}y^{-2}$

Next, use the reciprocal of the base raised to a positive exponent as follows: $\frac{1}{5^2 y^2}$

Finally, raise the number to the exponent to solve the problem as follows:

$$\frac{1}{25y^2}$$

PROBLEM # 23

SOLVE: $\frac{x^{10}}{x^4}$

CORRECT ANSWER: x^6

The above expression follows property (4) of the exponent rules which is explained in the front of this workbook.

Subtract the exponents on the variable as follows to solve the problem:

$$x^{10-4} = x^6$$

PROBLEM # 24

SOLVE: $\frac{\left(x^0\right)^5}{x^3}$

CORRECT ANSWER: $\frac{1}{x^3}$

A base raised to the power of **(0)** equals **(1)** therefore the (x) in the numerator equals to **(1)**. The problem is rewritten as: $\frac{(1)^5}{x^3} = \frac{1^5}{x^3}$

A base of **(1)** raised to any power equals itself. Therefore, the answer is as follows: $\frac{1}{x^3}$

SOLVE: $\dfrac{12r^{-6}s^0t^{-3}}{3\,r^{-4}s^{-3}t^{-5}}$

CORRECT ANSWER: $\dfrac{4\,s^3t^2}{r^2}$

The above expression follows property (4) of the exponent rules which are explained in the front of this workbook.

And also:

A base raised to a negative exponent will always result in the reciprocal of the base raise to the positive exponent.

And also:

A base raised to the power of **(0)** equals **(1)**.

Initially simplify the numbers by using the division operation as follows:

$12 \div 3 = 4$

Next follow property (4) and rewrite the problem in a simplified manner as follows: $4r^{-6-(-4)}1s^{-3}t^{-3-(-5)} = 4r^{-6+4}1s^{-3}t^{-3+5} = 4r^{-2}1s^{-3}t^2$

Finally, use the reciprocal of the base raised to a positive exponent and solve the problem as follows: $\dfrac{4\,s^3t^2}{r^2}$

PROBLEM # 26

SOLVE: $\dfrac{\left(3x^{-2}y^8\right)^4}{\left(9x^4\,y^{-3}\right)^2}$

CORRECT ANSWER: $\dfrac{y^{38}}{x^{16}}$

The above expression follows properties (2) and (4) of the exponent rules which are explained in the front of this workbook.

And also:

A base raised to a negative exponent will always result in the reciprocal of the base raise to the positive exponent.

Initially follow property (2) and simplify the problem by raising what is inside the parenthesis to the respected exponent as follows: $\dfrac{3^4 x^{-2\cdot4} y^{8\cdot4}}{9^2 x^{4\cdot2} y^{-3\cdot2}}$

Next, simplify the above expression by raising each number to its appropriate exponent as follows: $\dfrac{81 x^{-8} y^{32}}{81 x^8\ y^{-6}}$

Next, simplify further by dividing the number 81 into itself and follow property (4) and subtract the exponents on the (y) variable as follows:

$x^{-8-8} y^{32-(-6)} = x^{-16} y^{32+6} = x^{-16} y^{38}$

Finally, use the reciprocal of the base (x) raised to a positive exponent to complete the problem as follows: $\dfrac{y^{38}}{x^{16}}$

PROBLEM # 27

SOLVE: $(3x^2)(2x^3)(5x^4)$

CORRECT ANSWER: $30x^9$

The above expression follows property (1) of the exponent rules which are explained in the front of this workbook.

In order to solve the problem, follow property (1) and multiply the numbers and multiply the variables while adding their exponents as follows:

$(3)(2)(5)x^{2+3+4} = 30x^9$

SOLVE: $\left(\dfrac{8x^2}{4x^4\,y^{-3}}\right)^4$

CORRECT ANSWER: $\dfrac{16y^{12}}{x^8}$

The above expression follows properties (2) and (4) of the exponent rules which are explained in the front of this workbook.

And also:

A base raised to a negative exponent will always result in the reciprocal of the base raise to the positive exponent.

Initially follow property (2) and simplify the problem by raising what is inside the parenthesis to the respected exponent as follows: $\dfrac{8^4 x^{2\cdot4}}{4^4 x^{4\cdot4}\,y^{-3\cdot4}}$

Next raise the numbers to the exponents as follows: $\dfrac{4096x^8}{256x^{16}y^{-12}}$

Next simplify the numbers by using the division operation as follows:

$4096 \div 256 = 16$

Next follow property (4). Subtract the exponents on the (x) variable as follows: $\dfrac{16x^{8-16}}{y^{-12}} = \dfrac{16x^{-8}}{y^{-12}}$

Finally use the reciprocal of the (x) and the (y) variables raised to a positive exponent to complete the problems as follows: $\dfrac{16y^{12}}{x^8}$

PROBLEM # 29

SOLVE: $\dfrac{2^{-5}}{2^3}$

CORRECT ANSWER: $\dfrac{1}{2^8}$

The above expression follows property (4) of the exponent rules which are explained in the front of this workbook.

And also:

A base raised to a negative exponent will always result in the reciprocal of the base raise to the positive exponent.

Initially subtract the exponent on the base (2) in the problem as follows:

$$\frac{2^{-5}}{2^3} = 2^{-5-3} = 2^{-8}$$

Use the reciprocal of the base (2) with a positive exponent as follows: $\frac{1}{2^8}$

The solution to the problem can either be left as it is above or the base (2) can be raised to the exponent (8) if needed.

PROBLEM # 30

SOLVE: $\left(2x^2y^{-5}\right)^3\left(3x^{-4}y^2\right)^{-4}$

CORRECT ANSWER: $\dfrac{8x^{22}}{81y^{23}}$

The above expression follows properties (1) and (2) of the exponent rules which are explained in the front of this workbook.

And also:

A base raised to a negative exponent will always result in the reciprocal of the base raise to the positive exponent.

Initially follow property (2) and simplify the problem by raising what is inside the parenthesis to the respected exponent as follows:
$(2^3x^{2\cdot3}y^{-5\cdot3})(3^{-4}x^{-4\cdot-4}y^{2\cdot-4})$

Next use the reciprocal of the base (3) in the second parenthesis with a positive exponent as follows:

71

$$\left(8x^6y^{-15}\right)\left(\frac{1}{3^4}x^{16}y^{-8}\right) = \left(8x^6y^{-15}\right)\left(\frac{1}{81}x^{16}y^{-8}\right)$$

Multiply what is inside both parenthesis and add the exponents of the similar bases as follows: $\frac{8}{81}x^{6+16}y^{-15+(-8)} = \frac{8}{81}x^{22}y^{-23}$

Finally, use the reciprocal of the variable (y) raised to the positive exponent to complete the problem as follows: $\frac{8x^{22}}{81y^{23}}$

PROBLEM # 31

SOLVE: $\dfrac{t^{-10}}{t^{-4}}$

CORRECT ANSWER: $\dfrac{1}{t^6}$

The above expression follows property (4) of the exponent rules which are explained in the front of this workbook.

And also:

A base raised to a negative exponent will always result in the reciprocal of the base raise to the positive exponent.

Initially follow property (4) and subtract the exponents on the variable (t) as follows: $t^{-10-(-4)} = t^{-10+4} = t^{-6}$

Finally, use the reciprocal of the variable (t) raised to the positive exponent to complete the problem as follows: $\dfrac{1}{t^6}$

PROBLEM # 32

SOLVE: $\dfrac{(3x)^{-5}}{(3x)^{-8}}$

CORRECT ANSWER: $27x^3$

In order to solve this problem it is advisable to use positive exponents by using the reciprocal of the numerator and denominator. This will result in the numerator and denominators being switched as follows: $\dfrac{(3x)^8}{(3x)^5}$

Since the base is the same, the problem will then follow property (4) of the exponent rules which are explained in the front of this workbook.

Rewrite the problem as follows: $(3x)^{8-5} = (3x)^3$

Finally raise the base to the exponent as follows: $3^3 x^3 = 27x^3$

PROBLEM # 33

SOLVE: $(2x^4 y^{-3})(7x^{-8} y^5)$

CORRECT ANSWER: $\dfrac{14y^2}{x^4}$

The above expression follows properties (1) of the exponent rules which are explained in the front of this workbook.

And also:

A base raised to a negative exponent will always result in the reciprocal of the base raise to the positive exponent.

Initially follow property (1) by multiplying the numbers and adding the exponents on the similar variable bases as follows:

$(2)(7)x^{4+(-8)}y^{-3+5} = 14x^{-4}y^2$

Finally, use the reciprocal of the variable (x) raised to the positive exponent to complete the problem as follows: $\dfrac{14y^2}{x^4}$

PROBLEM # 34

SOLVE: $(2a^2b)^0$

CORRECT ANSWER: 1

A base (whether a number or a variable) raised to the power of **(0)** equals **(1)**. The answer to the above problems is as follows: $(2a^2b)^0 = 1$

PROBLEM # 35

SOLVE: $\dfrac{(2a^3b^{-2}c)^5}{(a^{-2}b^4c^{-3})^{-2}}$

CORRECT ANSWER: $\dfrac{32a^{11}}{b^2c}$

The above expression follows properties (2) and (4) of the exponent rules which are explained in the front of this workbook.

And also:

A base raised to a negative exponent will always result in the reciprocal of the base raise to the positive exponent.

Initially follow property (2) and simplify the problem by raising what is inside each parenthesis to the respected exponent as follows:

$$\frac{2^5a^{3\cdot5}b^{-2\cdot5}c^5}{a^{-2\cdot-2}b^{4\cdot-2}c^{-3\cdot-2}} = \frac{32a^{15}b^{-10}c^5}{a^4b^{-8}c^6}$$

Next follow property (4) and subtract the exponents on the variables as follows: $32a^{15-4}b^{-10-(-8)}c^{5-6} = 32a^{11}b^{-2}c^{-1}$

Finally, use the reciprocal of the bases **(b)** and **(c)** with a positive exponent to complete the problem: $\dfrac{32a^{11}}{b^2c}$

PROBLEM # 36

SOLVE: $\dfrac{\left(a^3\right)^2 \left(a^4\right)^5}{\left(a^5\right)^2}$

CORRECT ANSWER: a^{16}

The above expression follows properties (1), (2) and (4) of the exponent rules which are explained in the front of this workbook.

Initially follow property (2) and simplify the problem by raising what is inside each parenthesis to the respected exponent as follows: $\dfrac{a^{3\cdot2}a^{4\cdot5}}{a^{5\cdot2}} = \dfrac{a^6 a^{20}}{a^{10}}$

Next follow property (1) and add the exponents in the numerator as follows:

$\dfrac{a^{6+20}}{a^{10}} = \dfrac{a^{26}}{a^{10}}$

Finally follow property (4) and subtract the exponents in the numerator and denominator to solve the problem as follows: $a^{26-10} = a^{16}$

PROBLEM # 37

SOLVE: $2x^2y^3 \left(\dfrac{7xy^4}{14x^3y^6}\right)^{-2}$

CORRECT ANSWER: $8\,x^6\,y^7$

The above expression follows properties (1), (2) and (4) of the exponent rules which are explained in the front of this workbook.

And also:

A base raised to a negative exponent will always result in the reciprocal of the base raise to the positive exponent.

Hints to remember:

1. The first part of the problem $2x^2y^3$ is considered a numerator with a denominator of (1).

2. The second part of the problem $\left(\dfrac{7xy^4}{14x^3y^6}\right)^{-2}$ is raised to a negative power. It will be easier to tackle this problem by switching the numerator and denominator and changing the (2) exponent to a positive exponent as follows: $\left(\dfrac{14x^3y^6}{7xy^4}\right)^2$

The problem can now be rewritten as follows: $\left(\dfrac{2x^2y^3}{1}\right)\left(\dfrac{14x^3y^6}{7xy^4}\right)^2$

Next follow property (2) and simplify the problem by raising what is inside each parenthesis to the respected exponent as follows:

$$\left(\frac{2x^2y^3}{1}\right)\left(\frac{14^2x^{3\cdot2}y^{6\cdot2}}{7^2x^2\,y^{4\cdot2}}\right) = \left(\frac{2x^2y^3}{1}\right)\left(\frac{196x^6y^{12}}{49x^2\,y^8}\right)$$

Follow property (1) and multiply the numerators and denominator while adding the exponents on the similar bases as follows:

$$\left(\frac{392x^{2+6}y^{3+12}}{49x^2\,y^8}\right) = \left(\frac{392x^8y^{15}}{49x^2\,y^8}\right)$$

Follow property (4) and divide the numerators and denominator while subtracting the exponents on the similar bases as follows:
$$8\,x^{8-2}\,y^{15-8} = 8\,x^6\,y^7$$

PROBLEM # 38

SOLVE: $\dfrac{\left(x^{-7}\right)^3\left(x^4\right)^5}{\left(x^3\right)^2\left(x^{-1}\right)^8}$

CORRECT ANSWER: x

The above expression follows properties (1), (2) and (4) of the exponent rules which are explained in the front of this workbook.

And also

A base raised to a negative exponent will always result in the reciprocal of the base raise to the positive exponent.

Initially follow property (2) and simplify the problem by raising what is inside each parenthesis to the respected exponent as follows:

$$\frac{(x^{-7 \cdot 3})(x^{4 \cdot 5})}{(x^{3 \cdot 2})(x^{-1 \cdot 8})} = \frac{(x^{-21})(x^{20})}{(x^6)(x^{-8})}$$

Follow properties (1) and multiply the numerators and denominator while adding the exponents on the similar bases as follows: $\dfrac{x^{-21+20}}{x^{6+(-8)}} = \dfrac{x^{-1}}{x^{-2}}$

Follow properties (4) and divide the numerators and denominator while subtracting the exponents on the similar bases as follows:
$x^{-1-(-2)} = x^{-1+2} \quad = x^1 \quad = x$

PROBLEM # 39

SOLVE: $\left(\dfrac{x^{-5}y^2}{x^{-3}y^5} \right)^{-2}$

CORRECT ANSWER: $x^4 y^6$

The above expression follows properties (2) and (4) of the exponent rules which are explained in the front of this workbook.

Initially follow property (2) and simplify the problem by raising what is inside each parenthesis to the respected exponent as follows:

$$\left(\frac{x^{-5 \cdot -2} y^{2 \cdot -2}}{x^{-3 \cdot -2} y^{5 \cdot -2}} \right) = \left(\frac{x^{10} y^{-4}}{x^6 y^{-10}} \right)$$

Follow properties (4) and divide the numerators and denominator while subtracting the exponents on the similar bases as follows:

$$x^{10-6}y^{-4-(-10)} = x^4y^{-4+10} = x^4y^6$$

PROBLEM # 40

SOLVE: $\dfrac{\left(a^{-2}\right)^3\left(a^4\right)^2}{\left(a^{-3}\right)^{-2}}$

CORRECT ANSWER: $\dfrac{1}{a^4}$

The above expression follows properties (1), (2) and (4) of the exponent rules which are explained in the front of this workbook.

And also

A base raised to a negative exponent will always result in the reciprocal of the base raise to the positive exponent.

Initially follow property (2) and simplify the problem by raising what is inside each parenthesis to the respected exponent as follows:

$$\frac{\left(a^{-2\cdot3}\right)\left(a^{4\cdot2}\right)}{\left(a^{-3\cdot-2}\right)} = \frac{\left(a^{-6}\right)\left(a^8\right)}{\left(a^6\right)}$$

Follow property (1) and multiply the numerator while adding the exponents on the similar bases as follows: $\dfrac{a^{-6+8}}{a^6} = \dfrac{a^2}{a^6}$

Follow properties (4) and divide the numerators and denominator while subtracting the exponents on the similar bases as follows: $a^{2-6} = a^{-4}$

Finally, use the reciprocal of the base (a) with a positive exponent to solve the problem as follows: $\dfrac{1}{a^4}$

PROBLEM # 41

SOLVE: $8x^4y^{-3}\left(\dfrac{12x^{-3}y^{-2}}{24x^4y^{-5}}\right)^0$

CORRECT ANSWER: $\dfrac{8x^4}{y^3}$

A base (whether a number or a variable) raised to the power of **(0)** equals **(1)**. Therefore, the second part of the problem $\left(\dfrac{12x^{-3}y^{-2}}{24x^4y^{-5}}\right)^0$ will result in a solution of **(1)**.

The problem can now be rewritten as follows: $8x^4y^{-3}(1) = 8x^4y^{-3}$

A base raised to a negative exponent will always result in the reciprocal of the base raise to the positive exponent.

Use the reciprocal of the base (y) to solve the problem as follows: $\dfrac{8x^4}{y^3}$

PROBLEM # 42

SOLVE: $\left(\dfrac{ab^{-3}c^{-2}}{a^{-3}b^0c^{-5}}\right)^{-1}$

CORRECT ANSWER: $\dfrac{b^3}{a^4c^3}$

The above expression follows properties (2) and (4) of the exponent rules which are explained in the front of this workbook.

And Also:

A base raised to a negative exponent will always result in the reciprocal of the base raise to the positive exponent.

And Also:

A base (whether a number or a variable) raised to the power of **(0)** equals **(1)**.

Initially follow property (2) and simplify the problem by raising what is inside each parenthesis to the respected exponent.

The variable (b) in the denominator has been changed to **(1)** because it is raised to the power of **(0)**.

$$\left(\frac{a^{-1}\,b^{-3\cdot-1}c^{-2\cdot-1}}{a^{-3\cdot-1}(1)c^{-5\cdot-1}}\right) = \left(\frac{a^{-1}\,b^3c^2}{a^3(1)c^5}\right)$$

Follow properties (4) and divide the numerators and denominator while subtracting the exponents on the similar bases as follows:

$$a^{-1-3}b^3c^{2-5} = a^{-4}b^3c^{-3}$$

Finally, use the reciprocal of the bases (a) and $(c\ )$ with a positive exponent to solve the problem as follows: $\dfrac{b^3}{a^4c^3}$

PROBLEM # 43

SOLVE: $\quad \dfrac{5a^8b^3}{20a^5b^{-4}}$

CORRECT ANSWER: $\dfrac{a^3\,b^7}{4}$

The above expression follows properties (4) of the exponent rules which are explained in the front of this workbook.

Follow properties (4) and divide the numerators and denominator while subtracting the exponents on the similar bases as follows:

$$\frac{5}{20}\,a^{8-5}\,b^{3-(-4)} = \frac{5}{20}\,a^{8-5}\,b^{3+4} = \frac{1}{4}\,a^3\,b^7 = \frac{a^3\,b^7}{4}$$

SOLVE: $\left(\dfrac{x^{-3}y^2}{x^4y^{-5}}\right)^{-2}\left(\dfrac{x^{-4}y^2}{x^0y^2}\right)$

CORRECT ANSWER: $\dfrac{x^{10}}{y^{15}}$

The above expression follows properties (1), (2) and (4) of the exponent rules which is explained in the front of this workbook.

Also:

A base raised to a negative exponent will always result in the reciprocal of the base raise to the positive exponent.

Initially follow property (2) and simplify the problem by raising what is inside each parenthesis to the respected exponent.

The variable (x) in the denominator of the second fraction will be changed to (1) because it is raised to the power of (0):

$$\left(\dfrac{x^{-3\cdot-2}y^{2\cdot-2}}{x^{4\cdot-2}y^{-5\cdot-2}}\right)\left(\dfrac{x^{-4}y}{1y^2}\right) = \left(\dfrac{x^6y^{-4}}{x^{-8}y^{10}}\right)\left(\dfrac{x^{-4}y}{1y^2}\right)$$

Follow properties **(1)** and multiply the numerators and denominator while adding the exponents on the similar bases as follows:

$$\left(\dfrac{x^{6+(-4)}y^{-4+1}}{x^{-8}y^{10+2}}\right) = \left(\dfrac{x^{6-4}y^{-4+1}}{x^{-8}y^{10+2}}\right) = \left(\dfrac{x^2y^{-3}}{x^{-8}y^{12}}\right)$$

Follow properties (4) and divide the numerators and denominator while subtracting the exponents on the similar bases as follows:

$$x^{2-(-8)}y^{-3-12} = x^{2+8}y^{-3-12} = x^{10}y^{-15}$$

Finally, use the reciprocal of the base (y) with a positive exponent to solve the problem as follows: $\dfrac{x^{10}}{y^{15}}$

PROBLEM # 45

SOLVE: $\dfrac{\left(x^5\right)^6}{\left(x^3\right)^4}$

CORRECT ANSWER: x^{18}

The above expression follows properties (2) and (4) of the exponent rules which is explained in the front of this workbook.

Initially follow property (2) and simplify the problem by raising what is inside each parenthesis to the respected exponent.

$$\frac{x^{5 \cdot 6}}{x^{3 \cdot 4}} = \frac{x^{30}}{x^{12}}$$

Follow properties (4) and divide the numerators and denominator while subtracting the exponents on the similar bases to solve the problem as follows: $x^{30-12} = x^{18}$

ABOUT THE AUTHOR

Najwa Hirn holds a Bachelor of Science degree with honors in Engineering Technology. She has been working with Mathematics for over 25 years both professionally and privately. She taught math for many years.

Najwa is passionate about helping students succeed in Mathematics. She prides herself in being able to simplify math concept for students and teach every them according to their levels. Her step-by-step approach to solving problems has helped many students understand concepts better. She does not eliminate a step no matter how simple it may be since eliminating steps is what confuses many students.

Najwa Hirn can be reached at:

Learningmathquick@gmail.com

Made in the USA
Las Vegas, NV
08 February 2024